The Great Einstein Hoax

$$E \neq mc^2$$

$$R_{\mu\nu} - 1/2 R g_{\mu\nu} \neq \frac{8\pi G}{c^4} T_{\mu\nu}$$

outskirtspress

DENVER, COLORADO

Contents

About The Author

The author is not a physicist or in academia but works in a factory making industrial cleaners for Refil Chemicals Inc. He does not know a great amount of physics and none of the associated math. He is certainly not smarter or more learned than physicists but he is more skeptical and when things don't make sense or don't connect with reality it bothers him and causes a search for alternate explanations. If what you believe doesn't make sense you believe in nonsense. The fact that today's physics brags amount not believing in reason and creating worlds that have nothing to do with reality compelled him to write this book.

Foreword

The title of this book or article is misleading. Einstein was not trying to deceive or perpetrate a hoax he was trying to solve the problems of physics in his time. The solutions he proposed were wrong so a more appropriate title would be The Great Einstein Mistake but the continuing perpetuating and expanding of his errors by physicist is a hoax and it is this continuation that the title refers to.

Acknowledgements

I would like to thank Neil Genzlinger for his help in turning my fractured sentences into coherent English.

The Great Einstein Hoax

Introduction

EINSTEIN: THE NAME has become synonymous with genius, and he is one of the founding fathers of modern physics. Winner of the Nobel Prize in physics, creator of the photon, creator of the theory of special and general relativity, he is certainly the most famous and perhaps the greatest scientist of the last century and is regarded by many as the greatest scientist ever. But what if he was wrong about everything and all of modern physics is a sham? What if today's physics is just a modern version of "The Emperor's New Clothes" and in order to conform and be accepted physicists have abandoned reason and reality? The result of Einstein's work is that today physics has become fantasy where magic and invisible things are created to make observations conform to theories and where reason and logic are forsaken.

Scientists once scoffed at religious beliefs because they were based not on reason but on faith. Religions were considered a form of superstition that believed in an all-knowing entity, in miracles, magic, and invisible worlds of spirits. Today, physicists will admit that they do not believe in reason and do believe that most matter is invisible and inanimate objects know the future. They maintain that their job is not to make sense of the physical world but to describe it. This is comparable to a book reviewer who says that his job is to tell you what words are used in the book and how often they are used, not to tell you what the book is about or how good or bad it is. How can physicists scoff at an all-knowing being and then propose all-knowing electrons, ones that know if you are watching them or are going to be observing them in the future? The electrons not only know what their future path will be and all the fields they will encounter but also everything that will happen to you and what your future is. Physicists deny an all-knowing God but then create the omniscient electron.

They accept that it is possible for things to continuously disappear and instantly reappear somewhere else without moving from one location to the other, but they do not believe that this is magic. Einstein asserted that nothing can go faster than the speed of light, but for the electrons in the inner orbits of large atoms to have the momentum to balance the attractive force of the protons in the nucleus, either they must travel faster than the speed of light or the atoms must be much larger than they are. The solution for this dilemma was not to question Einstein's assertion about light but to have the electrons disappear and instantly reappear somewhere else. If something were to travel faster than the speed of light it would seem to disappear and then reappear somewhere else, so how does this disappearing act solve the problem? The area where the electron exists in an atom is defined by the probability of finding the electron in that area. These areas, or shells, are pear shaped for most electrons, and the electron can reappear anywhere within the shell; sometimes it is close to the nucleus and sometimes it is at a distance. If it is not moving and has no momentum, what counteracts the attractive force of the nucleus, and what produces the atom's magnetic field? What happens to the electrical field produced by the electron when it disappears and reappears? Does the electrical charge of the electron appear simultaneously with the electron, and what happens to other electrons as the negative charge of an electron disappears and reappears? These magical things happen to the electron because of the assertion by Einstein that nothing can go faster than the speed of light, and yet there are experiments and observations that show that things do go faster than the speed of light.

Because there is not enough observable matter to explain the theory of gravity and the actions of galaxies, physicist created invisible dark matter and dark energy. According to the latest calculations, 94 percent of the universe is invisible. This is from a science that is supposed to be based on reality and observable phenomena -- to maintain the validity of their theories, physicists have had to abandon reality and propose the imaginary. Experiments are not done to validate theories but to confirm them; any data that supports a theory is embraced no matter how dubious, and any data that disputes the theory is discarded. The negative result of the Michelson-Morley experiment is cited as proof of the constant speed of light, but all it really proves is that the theory of ether it was based on was wrong. Today the standard of proof supporting an accepted theory is low, while any evidence contradicting an established theory must survive the attacks of physic's magic spells. It is easy to prove what you already believe, and any confirmation is acceptable. A prime example of this is the experiment to detect the neutrino. The theory of special relativity does not give the right answer for the energy of radioactive particles emitted by an atom; to preserve the theory the neutrino, a subatomic particle that is undetectable, was created. An experiment was then devised to detect this undetectable particle, and not surprisingly it found the neutrino, except the neutrino found has nothing in common with the theoretical neutrino proposed. This shows that if you look for something long enough and hard enough you will find it whether it exists or not. A sure way to be awarded a Nobel Prize is to devise an experiment to detect some aberration that can be manipulated into evidence of gravitational waves. It has been almost a century since Einstein proposed the existence of gravitational waves, yet despite huge advances in technology and instrumentation no evidence of these waves has been detected. A Nobel Prize was awarded for the calculations of the existence of these waves, so any actual evidence of their existence, no matter how dubious, should also be awarded the prize just as the dubious detection of the neutrino was.

Another example of this search for evidence to validate theory is the particle accelerators or atom smashers that search for the subsubatomic particles that make up protons, electrons and neutrons. Particles are accelerated to close to the speed of light, and then crashed into other particles. The trails of the fragments are then examined to see if they support the existence of any subsubatomic particles proposed by theory. According to the uncertainty principle, scientists cannot know both the position and the velocity of the particles they are accelerating, but they can know how the tiny fragments of these particles behave. They use incredibly strong magnetic and electrical fields to conduct the experiments, but these fields are comparatively weak compared with the fields around the particles and atoms. How can you predict the effect of these fields and what happens to particles in the fields as they are changed? Accelerators smash thousands of atoms to find the trail of the shadow of a ghost of subatomic particles that will support physicists' theories. Modern physics is about creating illusions to support delusions.

This change in physics, from a science to the realm of metaphysics, is the direct results of Einstein's theories. The errors in classical physics that led to the acceptance of Einstein's theories can be explained by re-examining and reinterpreting the data because the theories of classical physics have a basis in physical reality. This is not the case with Einstein's theories. His theories are based on assertions of relativity and the constant speed of light that have no evidence to support them. His thought experiments turn out as expected because all thought experiments do, but experiments in the real world are rarely so compliant. It can be shown experimentally (Appendix 1) that the force between two magnets does not equal the product of the magnets over the distance squared and the force of a magnet does not decrease at approximately the cube of the distance. The strength of a magnet decreases linearly with the distance, and the force between two magnets is equal to the sum of the force of the magnets over the distance between the face of one magnet and the magnetic field of the other magnet. This error led Newton to use the same formula for the force of gravity, and this is why there are so many problems with the theory of gravity. Einstein's theory of gravity, general relativity, accepted the basis of Newton's theory, which was wrong, and then produced an incorrect theory that corrected the perceived flaw of the force of gravity traveling faster than the speed of light. Because the new theory did not correct the basic error in Newton's theory, physicists have proceeded to invent nonexistent entities -- dark matter, dark energy, black holes, the big bang theory -- to compensate for the failures or errors of Einstein's theory.

Einstein theories are not based on experimental data or observations of reality but on assertions, and as long as these assertions are accepted there is no way to disprove the theories. It is the same as someone believing that the Bible is literally true and if an inconsistency or evidence to the contrary is observed the problem must be with the observation because the Bible is the final authority and infallible. There is no way to convince a physicist that Einstein was wrong with evidence and logic because to counter these they have abandoned reason and developed the magical spells of the uncertainty principle and quantum physics. If there are experiments and observations that show things going faster than the speed of light, physicists can still believe that nothing can go faster than the speed of light because of their magic spells. At what point do things become so ridiculous and absurd that they question the unproven assertions? What evidence is necessary? Today with the belief that most of the matter and energy in the universe are invisible, the creation of a myriad of subatomic particles and the invention of multiple dimensions, physics has created a universe of make believe.

The difference between reality and make believe is that with reality there is physical evidence of the existence of something, while in make believe its existence is the result of a belief that it must exist according to accepted theory. Both the electrical force and the magnetic force are established in reality. The strong and weak nuclear forces are created forces necessitated by the theory of the structure of the atom. These created forces do not behave or have the same actions as the forces observed in reality, which is an indication that they are not real. The theory of gravity is based on falling object and orbiting satellites, but gravity's behavior is so unique that it also falls into the category of a created force. The challenge is to find alternative explanations or theories that can explain the phenomena that necessitated the creation of these forces or subatomic particles yet still conforms to the behavior of the rest of reality.

The purpose of this book is to return physics to the field of science and to have it once again be based on reason and reality. It will offer alternate theories or possibilities that are based on experimental data and observations that will replace Einstein's theories without the need to create new particles, forces or dimensions. For the disciples of Einstein, who believe in his infallibility, this will not be acceptable because for them the questioning of Einstein's theories constitutes heresy bordering on blasphemy. To show that something is wrong you must present an alternative that is more reasonable, simpler and better fits the data, but this is not possible for the true believers; for them, no matter how complicated and contradictory theories become they are accepted as gospel. If Einstein says you cannot do an experiment that distinguishes between a gravitational force and acceleration, then it can't be done -- not because it's not possible but because Einstein says that it cannot be done. His theory of equivalence between gravity and acceleration has several experiments able to distinguish between them, based on Einstein's own theories, but this does not deter the true believers. This book is not for these true believers; it is for those who believe in reason and experimental observations. To them it offers alternative theories that bring sanity back to physics. These theories and interpretations of data may not be correct, but at least they offer reasonable explanations rather than an imaginary one.

CHAPTER 1

Photons-Phoeyons

I HAVE A wonderful pet that I recommend to everyone. It is a fish-dog. When I want companionship and affection it is a dog, but when there is bad weather or I don't feel like taking the dog for a walk and cleaning up its droppings it is a fish. There is no need to fight to give it a bath; it becomes a fish. And instead of expensive dog food, which the dog throws up once a week, it eats fish food. No expensive grooming or vet bills, no need to worry that it might get loose and bite someone, no accidents on the carpet. The fish-dog provides all the benefits of dog ownership with the low cost maintenance and ease of care of a fish. I named my fish-dog photon because the type of animal it is at any given moment is determined by whatever I want it to be. Unfortunately the fish-dog, like the photon, does not exist.

Einstein's credibility was established by his theory of the photon. At the time, science was debating the nature of light and whether it is a particle or a wave. The experiments and observations supported the wave theory of light, with the exception of the photoelectric effect. When light strikes a bare metal or certain crystals, it causes an immediate electrical current, and the belief was that if light were a wave it would take time for the wave to transfer the energy necessary to dislodge an electron from an atom and create a current. It was also believed that increasing the intensity or amount of light striking the metal or crystal would also increase the current. Experimentation showed that this is not the case and that it is the wavelength or frequency of the light that causes the current; the intensity of the light is completely irrelevant.

Einstein's solution for this problem was the creation of the photon, a packet of waves with no mass that acted like a particle, dislodging the electron and causing the electric current. This theory correctly predicted that the current produced is a result of the frequency or wavelength of the light and not the intensity, but the photon theory did not clarify the nature of light. It just made it fuzzy, allowing scientists to treat light as either a wave or a particle, depending on which structure was convenient. These discrete packets of wave energy, or quanta, then gave rise to the quantum theory, which is one of the magic spells physicists use to explain why the impossible occurs. Einstein was awarded the Nobel Prize in physics for his photon theory and was transformed from a patent clerk into a famous physicist.

A particle and a wave are two completely different animals, and one cannot become the other. A particle carries energy as kinetic energy, while a wave is an energy disturbance within a medium. A moving hammer has kinetic energy equal to half the mass times its velocity squared. When the hammer strikes an object, it transfers energy to the object it hits. In the case of a solid object like a bell, the energy distorts the forces holding the metal in its shape. The forces return the shape to its original form, trying to re-establish equilibrium, and the energy is transmitted through the metal as a vibration, which then transmits the vibration and energy to the medium of the air around the bell, producing sound. If the hammer strikes the bell too hard and breaks it, the energy is transferred as kinetic energy to the pieces of the bell. Or if the hammer strikes something without a structure like a pile of sand, the energy is transferred as kinetic energy and no wave of energy disturbance is created. An object that has no mass, like the photon, has no kinetic energy and cannot transfer energy to an object like an electron or atom. The making of light into both a particle and a wave is just double talk, and the photon is a contrivance in which ignorance about the nature of light is given a name and called knowledge.

So if light has no kinetic energy, why does the photoelectric effect occur immediately when light strikes the surface? If light is an electromagnetic wave, why doesn't it take time to transfer enough energy to an atom to dislodge an electron? To understand the photoelectric effect it is necessary to know how the light is interacting with the material it strikes and how it is transmitted through these mediums. In the original experiment beams of light were directed at zinc metal plates, and this produced a current. The experiment worked only on bare metal, and if the zinc oxidized the photoelectric effect and the current ceased. To prevent the oxidation the metal was enclosed in a vacuum in a glass container, but this solution, while preventing the oxidation, also prevented the light from producing a current. It turned out that if the zinc was enclosed in a vacuum in a case made of crystal instead of glass, the photoelectric effect would again occur. Since glass transmits visible light but not ultraviolet light, while crystal transmits both visible and ultraviolet light, it was deduced that it was the ultraviolet light that was dislodging the electrons and causing the current.

Glass and crystal are chemically the same, so the difference in transmission of light must be the result of the different structures of glass and crystal. Glass is a liquid form of silica oxide, while crystal is the solid form. This would indicate that it is not the chemical nature of the material or atoms that determine what light waves are transmitted or reflected, but that it is a function of the structure the chemicals form. Diamonds and graphite are both made of carbon, but the structure of them makes the diamond transparent and the graphite opaque. They both absorb visible light; while the diamond emits this absorbed energy as visible light, the graphite emits it as other radiation. It is not the carbon atoms that are absorbing the light, but the structure they form. Colored crystals reflect certain wavelengths of light and transmit others, giving them their color. Different wavelengths of the electromagnetic spectrum are transmitted or reflected by different materials depending on their structures, not the atoms that make them, and it is the absorbed light that causes the photoelectric effect.

In the photoelectric effect, the light wave is not striking an electron or its atom to cause a current but is interacting with the magnetic and electric forces that form the structure of the medium. It is the distortion of these structures by the electromagnetic wave that dislodges the electron and causes a current, and not energy from the light being transferred to the atoms in the structure. If an electron in a structure is in an exposed position, the distortion of the structure by light of the right wavelength can knock it free, causing a current. This is the same principle as the piezoelectric effect, where the manual distortion of a crystal's structure dislodges an electron, causing an electrical current. For the wave to dislodge the electron it must be of the right wavelength or frequency to distort the bond that forms the crystal. If you wish to dislodge an object from a rope or cable, you hit it or pluck it, sending a wave through it and transmitting energy to the object. The size of the object determines the size of the wave needed to dislodge it. In the case of the photoelectric effect, the size of the wave, or wavelength, needed to dislodge the electron is dependent on the structure of the material. Ultraviolet light produces a current in zinc, while green light produces a current in photocells. The electromagnetic wave is distorting the structure of the medium, and this distortion is what dislodges the electron, causing the current, so there is no need for the time delay required to transfer energy to an atom.

By having light change the medium in which it is transmitted, the objections posed by the photoelectric effect to the wave nature of light are eliminated, and the photon becomes unnecessary. Light is an electromagnetic wave and not a particle, and as a wave its speed is determined by the medium in which it travels. When light moves from the air into glass, its speed changes because the forces forming the glass are different than those forming the atmosphere, and when it exits the glass and re-enters the air its speed returns to that of light in air. If light were a particle it would need to gain or lose kinetic energy as it passed through different mediums, creating changes in heat at the interfaces of the different structures. If a light wave were to travel through a multipaned window the light would give up energy creating heat as it slowed entering the first pane of glass. It would gain energy causing cooling as it left the glass and entered the gap between the panes. Repeating the process as it travels through multiple panes the light would produce cooling instead of warming on the inside of the window.

The creation of the photon established Einstein's reputation, but it caused serious difficulties to physics. The first problem it created was the precedent of solving problems of theory by creating subatomic particles that had no basis in reality. Today physics is populated with a myriad of such subatomic particles designed to fill any size hole in a theory and yet still be small enough to never be detected. It is easier to devise a subatomic particle than it is to modify or correct a theory. An invented subatomic particle can have any properties or behave in any way necessary to solve the problem, while modifying a theory is restricted by the constraints of reality. To preserve theory, physicists now not only create subatomic particles but also new dimensions, new realities, new universes and new forces whose only reason for being is to preserve an existing theory. This is the path that led physics into the present world of make believe, and it all began with Einstein and the creation of the photon.

The second problem with the creation of the photon is that it established the acceptance of duality. If light or an electromagnetic wave could have particle properties, then particles could also have wave properties. The wave nature of particles was established by one of the most famous or infamous experiments in physics, the dual thin slit experiment.

If you shine a light through two narrow slits that are close together, bright and dark bands will appear to be projected through the slits. This is an interference pattern resulting from the light waves reinforcing each other in some areas while canceling each other out in other areas. In the dual thin slit experiment, instead of light, electrons were directed at the two narrow slits, and this also resulted in the appearance of an interference pattern. This led to the conclusion that there was a both a particle and wave nature to the electron. When detectors were installed in the equipment to discover through which slit an electron particle passed, the interference pattern disappeared and a single spot was observed. From this the experimenters came to the conclusion that the electron knew when it was being observed and changed from a wave to a particle if it was being observed. To test this hypothesis they again modified the experiment so the detector would go on at random times and the observer would not know when the experiment was set to detect waves or particles. The results were that every time the detector came on, the interference pattern or wave nature of the electron disappeared and it converted to its particle nature. This led to the idea of the omniscient electron that not only knows if you are watching but also knows that you are going to be watching and changes its structure accordingly.

Particles do not have wave properties; what they have are electrical fields. (The electrical properties of a neutron will be discussed later.) As an electron moves, its negative charge disturbs the electric and magnetic fields that it travels through. This produces an electromagnetic wave, or light wave, that travels through the electric and magnetic fields. The speed of the electron is determined by the energy of the electron, while the speed of the light waves it produces is determined by the strength of the fields in which it travels. These two speeds are different, which results in the electron continuing to generate light waves as it moves through the fields. It is these light waves that create the interference pattern observed through the slits. When you detect the electron you either change its path or stop it, and this action prevents the generation of any additional light waves, causing the disappearance of the interference pattern. The electron has no wave properties; what it has is electrical properties that produce waves when it moves through an electromagnetic field.

To imbue an inanimate object with intelligence, emotion, or motive is not science but an attempt by the observer to humanize it, whether it's a teddy bear, a car or an electron. Electrons are not omniscient and do not care or know if you are watching them, and they do not know the spin of other electrons in an atom once they have left the electric and magnetic fields of the atom. The movement and actions of electrons are strictly the result of the fields they are in.

The third problem with the creation of the photon is its role in the development of quantum physics. Because subatomic particles did not appear to follow the observed laws of physics, physicists developed new laws governing the behavior of small particles. Whether an object is small or large is determined by the point of view of the observer and is not an absolute thing. The earth can be small or large depending on what it is being compared with, so to create special physics for one segment of reality is just a denial of either the existing laws of physics or the theories about the movement and behavior of small object. If you have multiple laws of physics, there must be an interface or graduation where there is a change from one set of rules to another. We do not accept separate laws of physic as things get larger, so why accept them for objects that are too small to be observed?

Quantum physics is a complicated and difficult subject, but it can be summarized thus: If you do the math wrong, you get the right answer. There is nothing new about this method; accountants have used it for centuries. But in science you are not supposed to have a "right" answer. You are supposed to discover the correct answer. If there is a conflict between what is observed and what is expected by theory, it is the theory that is wrong. What quantum physics does, in combination with the uncertainty principle, is allow any answer desired to be the right answer. Quantum physics and the uncertainty principle allow that a neutron can have more energy than a hydrogen atom, which would mean that as a star burns out after converting all its electrons and protons into neutrons and becomes a neutron star, it would contain more energy than when it was a star made up of hydrogen.

The quanta theory developed from the observation of adding energy to the electrons of an atom, which caused those electrons to release that energy as light. An electron does not gradually move from one orbit to a higher one as energy is added, it adsorbs energy and then jumps to a different orbit and emits light when releasing the energy and returning to the base orbit. This behavior of needing a quantum of energy to move to a higher orbit can be explained by a feedback mechanism. A moving electron produces a current, which in turn causes a magnetic field. The magnetic field, according to the right-hand rule, will push the electron toward the center of the atom. As the energy or speed of the electron increases, the current also increases, which then increases the strength of the magnetic field, preventing the electron from moving into a higher orbit. It is this feedback, where the increase in energy of the electron also increases the strength of the confinement field, which produces the quantum of energy that is necessary for the electron to move to a higher orbit and produce light.

The other part of physics magic spell that allows for solutions to problems that defy reason is the uncertainty principle. The Heisenberg uncertainty principle states that you cannot know both the velocity and the position of small objects such as electrons. The reasoning is that the act of observing the object will change either its velocity or its position. If you were to shine a light or other electromagnetic wave on an atom, molecule or electron, the electromagnetic wave would alter the position or velocity of the object, causing either its velocity or position to change, creating the uncertainty.

This assertion may or may not be true, and it is certainly not valid for larger objects. You can observe the effect a moving object has on other fields or objects and use those observations to deduce its velocity and position without affecting the object, but whether the principle is true or not is irrelevant. The problem with the uncertainty principle is that physicists have changed it from not knowing the position or velocity of an object to the assertion that the object does not have a particular position and velocity. If a friend is coming for a visit, you may not know precisely where he is or how fast he is moving, but this does not mean he does not have a position and velocity. By making their lack of knowledge into the "reality" that objects do not have a position and velocity, physicists have introduced ambiguity into everything, which allows for multiple possible answers.

The belief in the ambiguity of reality and in the idea that this is not a deficiency of the observer has led them to believe that reality is determined by the observer, and that for every possibility that exists there is a reality and it is the observer who makes or chooses what is real. The question used to be: If a tree falls in a forest and no one is there, does it make a sound? In today's physics this is a trick question, since if there is no physicist to observe the tree falling it cannot fall. Every time physicists close their eyes, the world ceases to exist. Physics is supposed to be an objective science, but this changing of the focus from the observation to the observer turns it into a subjective science like sociology. That this hypothesis of multiple realities is even discussed by physicists shows how delusional physics has become.

Einstein's photon did nothing to clarify the nature of light. What it did do was allow scientists to create subatomic particles as solutions for difficulties in theory, allow things to have two different natures where an object's properties are whatever is convenient to the theory, and invest reality with an uncertainty that makes everything possible. The credibility that the theory of the photon gave Einstein allowed him to make assertions on relativity and the constant speed of light that are the basis of his work. There is no evidence to support these assertions, and their acceptance has led physics to become divorced from reality and reason.

If the theory of the photon is wrong and light is an electromagnetic wave with no particle properties, then the speed of light cannot be constant but varies with the strength of the electrical and magnetic fields it travels in. And if the speed of light is not constant, then all of Einstein's theories are wrong.

CHAPTER 2

Unrelated Relativity

THE CONCEPT OF relativity comes from the discussion of whether two independent events occur simultaneously. Whether events occur simultaneously or not is purely from the point of view of the observer. If a listener is at the midpoint between two loudspeakers, the sound they emit will reach him simultaneously. But if the listener is not at the midpoint, the sound from one speaker will reach him first even though the speakers are broadcasting simultaneously. The problem exists not only for two things occurring but also for a single event. If a meteor were to strike the moon, observers could record the event as happening not only at different times but also on different days.

Einstein did a thought experiment to describe this. In his experiment a passenger on a train produces a flash of light just as the train passes a man standing at a station. The passenger sees the light striking the front and back of the railcar at the same time, while the man at the station sees the light strike the rear of the train car before it strikes the front of it. The motion of the train causes the back of the car to move toward the flash of light and at the same time causes the front of the car to move away from the flash. For the passenger moving with the train, these changes in distances are reversed when the light reflects from the two ends of the car and returns – he thus sees the light striking both ends simultaneously. For the observer at the station there is no compensating motion. This difference is a matter of perspective. If the passenger on the train were to put light sensors at the front and rear of the train car, he would see that the light actually strikes the rear of the car first and that his motion distorts his perspective. The problem for scientists is determining the correct perspective for making an observation or the correct reference point for a measurement. If a passenger in a different train that was moving faster than the one where the light flash occurred observed the flash, he would see the light strike the front of the train car first, and for him the other train would be moving in the opposite direction. There would be three different reference points and three different answers for where the light struck first.

If a group of physicist were observing the path of a bullet fired from a gun, they may debate what the correct reference frame for observation is, but they would all agree that the physicist in front who observed the bullet as a rotating disc growing in diameter had the wrong reference point. When thousands of people watch a race they each have a different perspective, but they agree that the correct reference point for determining who won the race is that of looking down the finish line. The problem for physicists is that objects have multiple motions and the observers are also in motion, so how is it possible to determine the correct reference frame? Einstein's solution was to make the speed of light constant and use it as the base reference point, and by doing this he made mass, distance and time variables. Instead of trying to determine the correct point of view, his solution to the problem was to make every point of view correct for its observer and to deny that there was a single correct reference point.

If several events were to occur in the United States at noon, these events would not be simultaneous but would be spaced hours apart because of the time zones. If they were to occur when the sun was at its highest point in the sky, which is reference noontime, they would occur at different times even if they were in the same time zone. If two events were to occur simultaneously in New York and Sydney, Australia, for the observer at each location they would be not only at different times but also in different seasons and on different days, and in the daytime for one and at night for the other. The answer each observer has for when something occurred is dependent on his perspective, and the question is what is the correct reference point or perspective to determine when something occurs or if events happen simultaneously. To solve the problem of different time references, different reference points were established. Time zones were created where the observers within the zone had the same time reference frames. For events that occurred in different time zones like New York and Sydney, or for events like a meteor strike on the moon that are observed in different times zones, a standard reference point was created: Greenwich time. But creating or determining an accepted reference point is not always easy.

Everyone observes the sun, moon, and stars rising in the east and moving across the sky to the west, and at one time everyone believed this was the correct perspective, with the earth being the center of the universe. That conclusion made perfect sense except that a few of the stars behaved oddly, sometimes even going backward in the sky. In today's physics this would not be a problem, since the few stars are not statistically significant and can be ignored. But in the past this bothered people, and they called these stars planets or wanderers. Copernicus came up with the solution that if the sun and not the earth were at the center of the universe, the motion of all the stars would make sense, as the moon orbited the earth, the earth and other planets orbited the sun and the stars orbited the sun. It was difficult for people to accept that even though the sun and moon appeared to follow the same path across the sky, the moon was orbiting the earth and the earth orbiting the sun. But because this perspective made sense of the motions of all the objects in the sky, after strong opposition by the established authorities it was accepted as the correct frame of reference. Today we know that the sun is not the center of the universe but of our solar system, which itself is orbiting in our galaxy, the Milky Way. Determining the correct perspective or reference point when considering all the various motions of the observers and the motions of the objects being observed is difficult, but to make all points of view correct is not the solution.

If you were to launch a rocket and put several satellites in orbit at the equator at different altitudes traveling to the east, you would observe that the satellite at an altitude of 36,000 km appeared to be stationary and that the satellites at higher altitudes seem to travel to the west. Since the satellite at 36,000 km does not appear to be moving and has no momentum to counter the force of gravity, it should fall to the earth, so there needs to be a reason that it remains in the sky. The obvious explanation is that there is a layer of subatomic particles, the contrareon, at the altitude where the satellite orbits that holds it aloft and also cause satellites above the layer of subatomic particles to move in the opposite direction. The correct explanation is that the observer is in motion and has the wrong reference point or perspective. If you were to use the center of the earth as a reference point, all the satellites would be orbiting to the east, with the satellite at 36,000 km moving at the same angular velocity as the observer on the surface of the earth. The satellites at higher altitudes, having a slower angular velocity, would appear from the surface to be moving backward.

The correct reference point for measuring the motion and position of an object is at the center of the energy field that controls the motion of the object. Energy fields form units like the earth, solar system or a galaxy, and the position, velocity and motion of objects in these fields is referenced to the center of the energy field forming the unit. Energy fields of units can then become a part of a larger energy field or unit, and the motion, velocity, and position of the subunits are then referenced to the center of this larger energy field. The moon has its own energy field where motion on it is referenced to the center of the moon. The earth is a separate unit, and motion, position, and velocities on it, including satellites like the moon unit, are referenced to the center of its energy field. The earth-moon unit is then part of a larger energy field that has the sun at its center. Objects on the moon and the earth have no velocity, position or motion relative to the sun but only to the energy field that controls their motion.

This theory of units, which says the correct reference point for observations of an object is the center of the organizing energy field in which it exists, is in conflict with Einstein's contention that light is the only valid reference point. To determine which theory is correct, it is necessary to examine the nature of light and time and determine if time is a variable and light is a constant as Einstein asserted.

CHAPTER 3

I See the Light

TO COME UP with his theory of special relativity, Einstein made an assertion that the speed of light was constant in a vacuum and nothing could exceed this speed. This was not submitted as a theory or hypothesis but was assumed as an indisputable fact. Einstein cited no evidence or data to support this assertion, and if the assertion is wrong all of his theories are wrong. Today physicists cite the Michelson-Morley experiment as proof that this assertion is true, but using the negative results of an experiment to prove anything is dubious science.

The Michelson-Morley experiment was never intended to test the speed of light but to detect the ether or medium in which a light wave traveled. The theory being that light, as a wave, would need a medium in which to propagate, and since light travels through the vacuum of space and in caves this medium called ether would need to permeate space and pass through the earth as it moved through space. The Michelson interferometer was designed to detect the motion of the earth as it moved through the ether in space. The path of the light traveling perpendicular to the earth's motion would be longer than the path of the light traveling parallel to the earth's motion, and the interference pattern created by these two light beams would change as the arms' position relative to the ether changed as the earth rotated.

The mechanism using light waves for measuring any change was sensitive enough to detect the difference in distance the light traveled in the ether, yet it detected no change in the interference pattern, and this was cited as proof that the speed of light is constant. The negative results of an experiment based on a theory known to be wrong are not proof of anything except that the theory is incorrect. If light is an electromagnetic wave, then the ether or medium in which it travels would be the electric and magnetic fields, not some artificial theoretical medium. And because the earth has its own electric and magnetic fields, the only way the experiment could be conducted correctly is at the interface of the earth's fields with those produced by the sun. Michelson was correct when he thought that the experiment's results showed that the ether was traveling with the earth.

Since Einstein's assertion was made over 100 years ago there has been no experiment showing that the speed of light is constant in a vacuum. There have been, however, experiments and observations showing things going faster than the speed of light, which is contrary to Einstein's assertion. Dr. Gunter Nimtz did an experiment in which microwaves traveled faster than the speed of light through a solid medium. To maintain Einstein's assertion and explain the results of this experiment and other experiments giving similar results, some physicists created a new subatomic particle, the tunneling photon. According to this explanation, photons drafted on each other, like racecars, allowing them to seem to go faster than the speed of light. This explanation demonstrates that those physicists are not fans of racing. Drafting in a racecar does not increase the maximum speed of the car; it reduces resistance, allowing the car to get closer to its maximum speed. If two cars are equal, with a maximum speed of 100 miles an hour, the car in front will be driving into the air, making it seem like it is driving into a head wind, which will reduce its speed to less than 100 miles an hour. A car following close behind will ride in the air behind the front car and avoid the head wind, allowing for a reserve of speed to pass the front car. Tunneling photons do not provide an explanation for why an electromagnetic wave is traveling faster than the speed of light; they are just an attempt to provide an excuse allowing for the continuation of the belief in Einstein's assertion. Astronomers with new telescopes have observed stars near the center of galaxies traveling faster than the speed of light, and, pending the creation of tunneling stars, there is no explanation for these.

It is possible to observe light going faster than the speed of light in a vacuum in a swimming pool. An image that we see consists of light waves traveling in unison where there is no interference between the waves, which would destroy the image. The farther we get from an object, the more neighboring light waves interfere, causing the image to become smaller and details to disappear. When looking at objects underwater everything appears closer because the light is traveling faster in the water and there is less interference between the light waves. If you look at an object floating on the water, the submerged part will look larger and closer than the section above the water because the light waves in the water are traveling faster than the light waves in the air.

lens

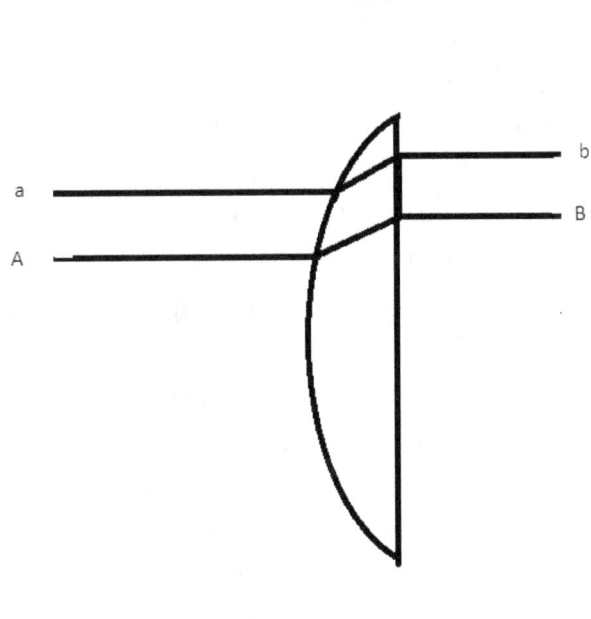

AB > ab

It is also possible to demonstrate this increase in the speed of light in a solid or liquid with a magnifying glass made with a simple lens. Two beams of synchronized light that are part of an image strike the curved face of the lens, one toward the center of the lens and one toward the edge. On striking the lens the waves refract and proceed at an angle to the back of the lens, where they exit the lens and un-re-fract, continuing their path. For the image to be preserved through the lens they must travel without interfering within the lens and exit the lens simultaneously. The light wave striking the lens toward the center travels a greater distance than the one striking near the edge, and the only way the light waves could travel without interfering with each other and exit the lens simultaneously is if the light travels faster in the lens than it does in the air.

It may seem strange to have something traveling faster in a solid or liquid than in a gas, but light is a wave, and the speed of the wave is determined by the medium in which it travels. Sound waves travel faster in water than in air because the molecules of water are bound together with a stronger force than the air molecules. Light is an electro-magnetic wave, and the stronger magnetic force holding the atoms or molecules of a solid or liquid together means the light travels faster in these mediums than in air. It is not traveling through the matter but through the electric and magnetic fields of the structure the matter forms. This is the same principle for the sound emitted by a guitar string. The tighter the tension on the string, the faster the wave travels and the higher the frequency of the sound emitted, so to tune the guitar the tension on the strings is adjusted. It doesn't matter whether the strings are made of steel, cop-per or catgut; the sound is a result of the vibration of the string and its tension.

It is the increased strength of the magnetic field around

the sun that causes light from distant stars to bend around it, not the warping of the space-time continuum that Einstein proposed in his general relativity theory. This change in wavelength and frequency of light due to a change in magnetic field also explains the red and blue shift of the spectra of light from distant stars and galaxies. It is not caused by the Doppler effect of an expanding universe; it is a result of the light traveling through magnetic fields of differing strength. The Doppler effect is caused when a wave traveling toward or away from an observer has a shift in wavelength and frequency. An object moving toward the observer will produce shorter waves, blue shift, while an object moving away from the observer will produce longer waves, red shift.

The effect is strictly a function of the motion between the object and the observer and has nothing to do with other motions of the observer or object. The object is either moving toward the observer or it is moving away, so how can the Doppler effect cause light from a distant star to have both a red and blue shift? As light travels into a weaker magnetic field, the wavelengths get longer and the frequencies decrease, causing the red shift. Then, if this light wave enters a stronger magnetic field, the wavelengths shorten and frequency increases, causing a blue shift -- giving the light both a red and a blue shift. The Zeeman and Stark effects are where the spectrum of light emitted by an atom changes in the presence of strong electric or magnetic fields, and this is the same thing that happens when light travels through changing fields in space. The shifts of electromagnetic waves are seen in experiments and equipment such as the MRI scanner, so it is surprising that they are not accounted for in light traveling through space where fields vary with the distance from stars. Attributing this red shift to a Doppler effect, as Einstein did, has led to the belief in a universe that is expanding at an accelerating rate and resulted in the creation of invisible matter and forces (dark matter an dark energy) to make this expansion conform to theory.

Physicists talk of being able to see back to the early stages of the Big Bang, but this is nonsense. Look at an object and then turn around. You can no longer see the object even though the light coming from the objects is in front of you. If you hold a mirror up you will again see the object, showing that the light from the object is also in front of you. You can only see light that is traveling toward you. If the Big Bang occurred billions of years ago, the light would be traveling away from the source, so where would you stand or in what direction would you look to have this light traveling toward you and to be able to observe it? Because of the time it takes light to travel a distance, all observations are of things that happened before they were observed, but this is not the same as looking back in time. Time is just a measurement of change, and in the next chapter the concept of time will be examined.

The assertions that the speed of light is constant and that nothing can exceed the speed of light in a vacuum are the basis of Einstein's theories of relativity and gravity, and if the assertions are not correct, then his theories are incorrect and modern physics is all make-believe. It is no wonder that physicists will believe in anything that preserves these assertions and ignore anything that contradicts them. It is easier to give up logic and reason than to admit that the physics is wrong.

CHAPTER 4

Time to Waste Time

THE SPEED OF light is not constant, and the only thing that is constant is change. Things change continuously at different rates, and this is what gives us the perception of time. Winter lasts for three months, ice ages last for eons, but everything changes, and time is how we quantify these changes. It is the memory of how things were that makes us aware of changes and allows us to project change into the future. Without a memory there would be no concept of time or the future; only the present would exist. This remembering of past occurrences gives a survival benefit because lessons learned can provide guidance when similar situations arise. Knowing when a rain is coming and where flooding occurs allows for action to be taken to minimize the hazards or take advantage of the opportunities that these events provide. The knowledge of changes that have occurred and the patterns of these changes are the basis of our belief in time.

The individual perception of time is purely subjective. Different individuals will believe a movie lasted forever or was over quickly depending on their involvement with the movie. This is subjective time. It is different for every individual and will even change for the individual. When a person sleeps or his attention is concentrated on a single subject, he has no concept of the passage of time because he does not note the changes that occur indicating the passage of time. This subjective time makes it hard to communicate to others information that increases group survival, and that led to the development of referral time.

Referral time is where individuals can communicate to others information based on common experiences and a common reference point. On earth the basic unit of referral time is the day, which is a reference to the rotational energy of the earth and represents one rotation of the earth. This unit of time is broken down into smaller units, which can represent degrees of change in the position of the sun or when the sun rises, is at its highest point or sets. These subdivisions provide a means for communicating a starting reference point and duration of time between individuals. Being able to communicate how long one needs to travel to find water or food is a survival benefit.

The next unit of referral time is the month, which is a reference to the orbital energy of the moon. In today's calendar the lengths of the months have been altered from purely reflecting the orbital energy of the moon to make each month have a whole number of days and each solar year have a whole number of months. This was done because the orbital energy of the moon is independent of both the rotational energy of the earth and the orbital energy of the earth, so it is more convenient to have a month start at the beginning of a new day and a year start at the beginning of a new month.

The final unit of referral time is the year, which is the time it takes the earth to make one rotation around the sun and is referenced to the orbital energy of the earth. In some cultures and calendars the lunar year is used, where the length of a year is related to orbits of the moon and is not in agreement with the solar calendar.

All of these reference units of time are based on energies associated with the earth and peculiar to the earth. On Venus, where the rotational energy and orbital energy of the planet are the same, one day is the same length as one year, and since there is no moon there is no unit of a month. Since referral time is different on every planet and in every solar system, it indicates that time is not a basic unit of physics. The charge of an electron, the makeup of a hydrogen atom and the chemical reactions of elements are the same throughout the universe and are basic to physics, while time is an arbitrary thing peculiar to a particular situation. Time is an invention to provide a means of communicating, just as the meter is an arbitrary unit invented to communicate distance. The meter, liter and gram are means to quantify size, volume and weight, while time is means of quantifying change and the rate of change. The subdividing of time into weeks, hours, minutes and seconds is purely arbitrary and is done for convenience. The day could consist of 10 hours that contain 100 minutes made up of 100 seconds if it were decided that metric time were more convenient. It is easier to say that something took one second than that it took .0011574 days, even though they both represent the same amount of time. Years, hours, days and even eons can all represent a long or short amount of time depending on what is being referenced.

Reference time can refer to things other than the energy associated with the earth. The term dog years does not mean the earth is orbiting the sun seven times faster for a dog. It is a means of equating the lifespan of a dog with that of a human. It does not equate the rate of change that a dog experiences with that of a human -- a year-old dog is full grown, while a 7-year-old human is not even half grown -- but the just the lifespan. This is what time is: a way of relating the changes in other objects to the rate of change we experience. It is simply a means of quantifying change, and it is no more real than the meter, gram or liter, which are the means we use for quantifying other measurements.

Reference time is artificial, but the creation of clock time is like creating imitation plastic. It is an approximation of reference time originally created to facilitate railroad schedules. With different localities setting their clocks to noon when the sun was at its apex, arrival and departure times at different stations were erratic. To correct this problem, time zones were created, and clocks within the zones were set to the same time instead of referral time. The zones are approximately one hour in width, but the borders are adjusted for political reasons and tend to follow political borders. These borders are where it is possible to travel back in time or into the future by stepping across a line. This is not real time travel but a result of the creation of clock time. When scientists talk of time slowing down as velocity increases, they are talking about clock time. This makes it possible, according to them, to purchase a device that will slow the aging process: just buy a watch that runs slow. This is the same as losing weight by adjusting the scale to register light and has nothing to do with reality.

The conversion of time into a dimension, creating the space-time continuum, is another artificial development in the treatment of time. Scientists say that by adding time as a fourth dimension they can describe the exact position of an object. Describing time as a dimension implies that time travel into the past or future is possible. Time travel makes for interesting science fiction stories, but in reality it is impossible. If a person were to build a machine and travel back in time, the energy and matter that make up the machine and its occupant would already exist in some other form. Traveling back in time either would result in the creation of new matter and energy or the energy and matter that make up the machine and occupant would disappear from the existing form. Jumping forward in time would involve the disappearance of energy and matter. Both circumstances would violate the principle of the conservation of energy and matter. All physical things change in both energy and matter over time. They change at different rates, but change is constant, and to believe an object that existed in the past is the same object as the one in the present or future defies the basis of time. There is no time dimension, just as there is no such thing as time.

CHAPTER 5

Not So Special Relativity

THE FORMULA E equals MC squared was proposed to explain the energy of radioactive particles emitted from the nucleus of atoms. With the conservation of energy, where matter and energy cannot be created or destroyed, the question was, where would the energy come from to overcome the electrical and binding forces that hold a particle in the nucleus of an atom and then expel it from the nucleus with such energy? The explanation was that part of the mass of the nucleus was converted into energy by some unknown means, and this energy then propelled the particle from the nucleus.

The solution proposed stated that matter and energy were different forms of the same thing, just like ice and steam, and could change from one form to the other. Why this conversion would be a function of the speed of light when light has nothing to do with the conversion is a complete mystery. Asserting that the speed of light is constant and converting it to an energy unit by squaring it produced a huge amount of energy from a minuscule amount of matter, making the actual measurement of mass change impossible. The selection of the speed of light in a vacuum for an event occurring in the nucleus of an atom, where matter is at its densest, seems rather strange. Why not say energy equals the mass times Avogadro's number and an energy constant with bizarre units, just as the gravity constant has completely nonsensical units? Avogadro's number would be large enough to provide massive amounts of energy from only a little matter, and by inventing a constant the value and units could be adjusted to give the right answer. The formula energy equals mass times the speed of light squared lacks the right units and says that energy is the same as kinetic energy.

A problem developed with Einstein's theory when the calculations were done comparing the energy of the emitted particles from a radioactive atom to the amount of energy expected from the missing mass of the atom -- it was discovered that the observed energy was insufficient to account for the missing mass. In most disciplines when a formula does not give the correct answer the formula is considered to be wrong, but not in physics. To maintain the correctness of the formula and explain why no other particles or electromagnetic waves were detected, a new subatomic particle, the neutrino, was created that contained the missing energy. The neutrino had no mass and no charge and traveled at the speed of light. And because it has none of the properties of matter, there is nothing to interact with the matter or fields of atoms. It was designed and thought to be undetectable. Like all good invented subsubatomic particles, it was big enough to fill the hole in the theory and yet small enough never to be detected.

The neutrino was designed to be undetectable, so an experiment was devised to detect the neutrinos emitted from the interior of the sun by nuclear reactions. A tank was buried 1500 m deep in the earth to filter out cosmic radiation and then was filled with 10,000 gallons of carbon tetrachloride. If a neutrino were to strike the nucleus of a chlorine atom, it would convert the chlorine into a radioactive argon atom, and the radiation of the argon atom could then be detected. Each month the tank was flushed with helium to separate out any argon atoms and then tested for their presence. It detected 10 radioactive argon atoms per month. It also found 10 atoms of non-radioactive argon for which there was no explanation, and so these atoms were ignored. The results of the experiment were cited as confirmation of the existence of the neutrino, but the results posed a new problem in that the number of neutrinos detected was only one-third that expected from the nuclear reactions occurring in the sun.

Another experiment using a different detection method was conducted in Japan at a depth of 1000 m, and this too detected the neutrinos, but this time at half the expected number. If a change in depth of 500 m changed the number of neutrinos detected from 15 a month to 10 a month, the question should arise of how the neutrinos ever got out of the interior of the sun to begin with, but with these experiments the existence of the neutrino was established, and details or explanations could be developed later.

To compensate for the missing neutrinos, it was proposed that there were three types of neutrinos, of which only one could interact with the nucleus and be detected. The neutrinos could change from one type to another, and therefore the number of neutrinos detected equaled the number expected and the experiment confirmed Einstein's theory. The only problem was that if the neutrinos traveled at the speed of light, according to Einstein's theory of relativity, there would be no time, and therefore the different types of neutrinos could not change from one type to another. To rectify this new problem, the nature of the neutrino was changed and it was postulated that the neutrinos did not travel at the speed of light after all. This allowed them to change from one type to another. This solution to the problem of course resulted in a new problem, because the change meant that the neutrino had to have some mass. According to Einstein's relativity theory, objects with no mass must travel at the speed of light and things with mass cannot travel at the speed of light.

The result is that an experiment designed to detect an undetectable particle with no charge or mass traveling at the speed of light comes to the conclusion that the particle has mass, is not traveling at the speed of light and has two other undetectable forms. This presents a new problem in that if the neutrino now has mass, then some missing mass in the nuclear reaction is not converted into energy and the expected number of neutrinos emitted by the nuclear reactions of the sun is reduced. It is now necessary to redo all the manipulations that have been done to experiments' results to make them conform to the old expectations and make them instead conform to the new expectations in order to prove the existence of a neutrino that has nothing in common with the one originally proposed.

The design and execution of these experiments are fundamentally flawed. Putting an experiment underground to hide it from radiation is like burying something to keep it from getting dirty. There are radioactive isotopes in the earth, air, materials used to make the experiment and the people conducting the experiment, so there is no way to isolate the experiment from all radioactive isotopes and thus detect only neutrinos emitted from the sun. The results of the experiment, 10 atoms in a month, are too precise given all the factors that could contribute to a margin of error. The ignoring of the non-radioactive argon atoms, the lack of any control standard as a comparison and the need for a 200 percent fudge factor to conform to expectations render the results meaningless. The results of the experiment do not support the creation of a neutrino fudge factor or the theory it was designed to preserve. The acceptance of the validity of the experiment and the fact that it was awarded a Nobel Prize show the lengths to which physicists will go to preserve their make-believe world.

The attraction of the formula E equals MC squared has a lot to do with the poetic rhythm of it and its simplicity. If you were to say energy equals two times the kinetic energy of light it would sound ridiculous, but you would be saying exactly the same thing. When you square velocity it becomes a unit of energy, and you can no more manipulate that unit into time or distance factors than you can get a distance or weight factor out of a torque unit. You cannot determine the length or weight of a bolt or screw from the kilogram-meter unit of torque applied to it, and you cannot separate distance and time from an energy unit.

The equating of mass and energy leads to contradictions and impossibilities. If an object were caught in the gravitational field of a black hole, it would accelerate, and as it accelerated it would increase in mass and energy. This increase in mass would increase the gravitational attraction between the black hole and the object and so increase the force between them. (If you have a set mass and divide it into two parts, the maximum gravitational force is when the two parts have equal mass.) Since energy cannot be created or destroyed, the black hole must be transferring mass to the object, and a corresponding decrease in the mass of the black hole must occur. All the objects caught in the gravitational pull of the black hole will have mass transferred to them, and eventually the black hole will lose enough mass and cease to be a black hole. If this were the case, then before a sun collapsed to form a black hole the same gravitational force would transfer mass to objects caught in its gravitational field and prevent the formation of the black hole.

Energy and matter are two completely different things; energy cannot become matter and matter cannot become energy as Einstein's theory states. The development of nuclear energy may seem to contradict this, but according to the theory of nuclear fission and fusion it is possible to split a helium nucleus into two deuterium nuclei and release energy and then use this energy to fuse two deuterium nuclei into a helium nucleus, releasing even more energy. This is clearly impossible, since it results in the creation of energy, violating the conservation principle, and so the theory of nuclear fission or the theory of fusion is incorrect. Or both are.

Energy and matter are two separate things that combine to form atoms. The matter gives the atom substance, while the energy gives it structure. A hydrogen atom is a neutron with added energy. The base state of matter is the neutron molecule, where a proton and an electron are stuck together by an electrical force. The electrical properties of matter give an atom its electrical fields, while the energy gives it a magnetic field or energy field. The fields of atoms then combine to build larger and larger objects, and it is the electrical and magnetic fields of objects that give them their properties. The size of an atom is not defined by the orbit of its outermost electron or matter but by the size of the fields of the atom. The size of the earth is not defined by the matter it is made of, but by the fields produced by the matter and energy that form it. To understand how objects behave and their properties, it is necessary to understand how energy and matter combine to form atoms and produce properties and fields.

CHAPTER 6

The Atom

THE CURRENT THEORY of the atom has a nucleus consisting of protons and neutrons surrounded by electrons in various shaped shells. This nuclear nature of the atom was established by experimentation in which alpha particles were directed at a thin sheet of gold. Most of the particles passed through the gold sheet, but a few were reflected back toward the source, which led to the conclusion that most of an atom is empty space and that the mass and positive charges of the gold atom were concentrated in a small nucleus, causing the reflection of the alpha particles. These conclusions posed the question of how the nucleus could hold together when the positive charges of the protons produce a repelling electrical force that should destroy it. Also, why should the neutrons, which have mass but no charge, be concentrated in the nucleus of the atom?

The strong nuclear force was created to hold the nucleus together, counteracting the repelling force of the protons. When this force was found to be inadequate, the weak nuclear force was created as a separate force to explain why the nuclei of most atoms were stable. These forces, like most created solutions, did not behave like the forces observed in nature. The forces are more powerful than the electrical force of the protons and must increase exponentially with the increase number of protons in the nucleus. The range of the nuclear forces is limited to the size of the nucleus and does not decrease as a function of distance as the other forces do. If the nucleus of an atom is a sphere, the protons on the surface will experience the greatest repelling force, pushing them out of the nucleus, necessitating a larger binding force than the protons in the center. The relation of the neutron to the nuclear forces is unclear. You would expect that the more neutrons there were in a nucleus, the larger the separation of the protons, the weaker the electrical repelling force, and the nucleus would become more stable. This is not the case, since if there are too many or too few neutrons in the nucleus it becomes unstable, and the binding force is not strong enough to hold it together, turning it into a radioactive atom emitting particles. It is necessary to examine the nature of the neutron to understand the nucleus of an atom. Why is it that a neutron is stable for millions of years when it is in the nucleus of an atom, but once out of the atom it decomposes into a proton, electron, and gamma ray within a half hour?

The neutron is thought to be a particle having no charge and with the combined mass of a proton and an electron. The experiments to determine the nature of radioactive particles that make up the nucleus of the atom were conducted by putting different radioactive isotopes in a lead container and allowing a beam of particles to escape through a small opening. These particles were directed between two plates, one with a positive charge and one with a negative charge. The particles that were attracted to the positively charged plate were called beta particles and later discovered to be electrons. The particles that passed between the two plates without being attracted to either plate became the neutron, and the particles attracted to the negative plate were called alpha particles. The mass and charge the alpha particle indicated that it was a subatomic molecule composed of two protons and two neutrons. This posed a problem. How would neutrons stick the protons together, counteracting their electrical repelling force and forming a stable structure like the alpha particle, while the neutrons themselves were unstable when out of the nucleus? The solution to this problem was to create a subsubatomic particle, the gluon, which jumped between the proton and neutron, gluing them together and forming the stable alpha particle. Why the gluon would jump from the neutron to the proton and back and why this would hold the two together are not explained, but the nice thing about invented subsubatomic particles is that they can behave any way necessary and perform any action necessary and nobody will be able to detect any evidence to dispute the claims.

Why would a neutron with no electrical charge be unstable while the alpha particle, with repelling electrical charges, is a stable structure? An alternate theory is that the neutron, like the alpha particle, is a subatomic molecule made up of a proton and an electron and has both a positive and a negative charge. This molecule would satisfy the experimental results of passing between two charged plates without being attracted to either plate. And if the neutron had a negative charge, it would adhere to a proton, acting as glue without necessitating the creation of the gluon.

All of matter has charges and electrical fields. When we say something is neutral it does not mean there is no charge but that the different charges mask each other preventing us from detecting them by electrical means. An atom normally has an equal number of electrons and protons which have electrical charges and fields within the atom. These fields hide each other making the atom appear to be neutral when in reality it consists of multiple charged particles and fields. This is also true of the neutron and all objects made of matter. The earth has both a negative field and a positive field which is demonstrated by the Van Allen belts which are made of charged particles surrounding the earth.

This interpreation of a dual charge of a neutron would also explain one of the contradictory problems of the neutron. When a proton and electron combine to form a neutron, it is an exothermic reaction, giving off energy, and yet when a neutron spontaneously splits into a proton, electron and gamma ray, this is also an exothermic reaction. The formation and destruction of the neutron cannot both be exothermic, since this again would involve the creation of energy and violate the conservation of energy, so something else must be providing the energy for one of these reactions. If a neutron had both a negative and a positive charge, as it moved it would represent two currents going in opposite directions. If this movement were through a magnetic field, the magnetic field would push the two currents in opposite directions, creating a shearing force that would split the neutron. The magnetic field would then provide the energy necessary to split the neutron into an electron and proton, prevent them from recombining and create a gamma ray, and so maintain the conservation of energy. Making the neutron a molecule with both a positive and a negative charge would eliminate the gluon and explain why the neutron is unstable when not in an atom's nucleus. The nucleus of an atom would then consist of protons and electrons, with the electrons acting as glue helping to hold the structure together, and the alpha particle, which is the nucleus of a helium atom, would consists of four protons and two electrons. While the electrons in the nucleus would help stabilize its structure, the nucleus would still contain more protons than electrons and would be electrically unstable, so the need for a strong nuclear force to hold it together remains.

The nucleus of the atom is surrounded be orbiting electrons that make the atom electrically neutral. The current theory on these shells of electrons is that they have three shapes: s shells are circular, p shells are pear shaped along the perpendicular axis of the atom, and d shells are pear shaped along the diagonals of the atom. These shapes and positions were determined by finding the probability of locating an electron in a particular area. The problem with the shapes obtained is that these probabilities were calculated using an assumption that the atom does not exist in a magnetic field. In the presence of a magnetic field, an electron will take the path perpendicular to the magnetic field, and instead of forming pear-shaped shells the electrons will form orbits. This theoretical atom with its p and d shells does not exist on the earth, in the solar system or anywhere in the observable universe where a magnetic field exists. If you start your model with an assumption that you know is incorrect, you cannot expect to get an accurate model. In the presence of a magnetic field the electrons of an atom form orbits and resemble the solar system. And with this shape it, although electrically neutral, will have both magnetic and electrical directional forces.

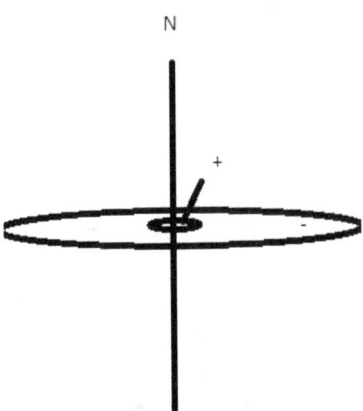

The electrical and magnetic forces would be directed along the north south axis of the atom, so the question arises: Why don't atoms stick together magnetically along this axis, creating molecules consisting of long strings of atoms? The reason molecules are not a string is that if the magnetic attractive forces were to align, the repelling electrical forces between the atoms would also be aligned. The positive charges of the nuclei would face each other, as would the negative forces of the orbiting electrons. A more stable structure would be formed if the atoms were offset, forming a lattice where the north pole of one atom was attracted to the four south poles of neighboring atoms and the repelling electrical forces of the atoms were not aligned. Picture a cube with an atom at each corner and an atom in the center. The north pole of the center atom would be attracted to the south poles the atoms in the corners at the top of the cube, while the south pole of the atom would be attracted to the north poles of the atoms in the corners at the bottom of the cube. The repelling force of the electron discs would be offset and reduced, making for a more stable structure than strings of atoms.

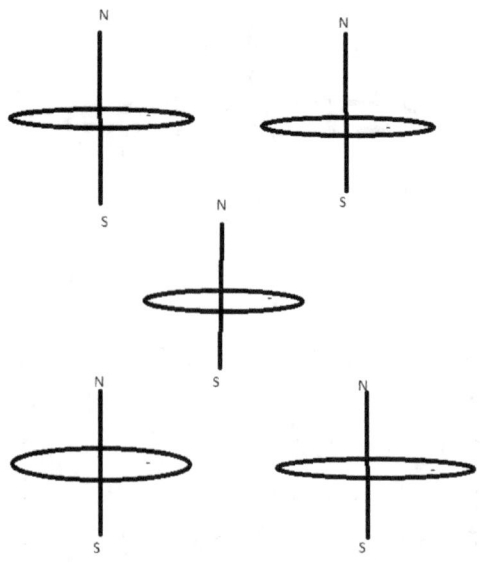

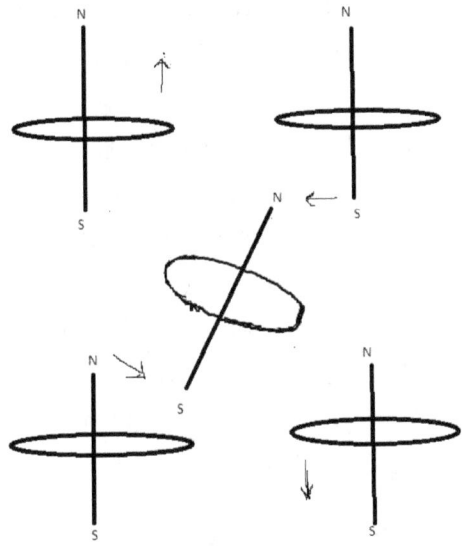

A structure such as this would also make the ideal medium for the transmission of electromagnetic waves. An oscillation in the center atom would be transmitted to the neighboring atoms by either an increase in the attraction between the magnetic poles or an increase in repelling force between the electron discs. When the magnetic poles move closer to the poles in the neighboring atoms, it increases the attractive force, transmitting the motion to those four atoms, while at the same time the movement of the electron discs closer to the electron discs of the four neighboring atoms increases the repelling force and transmits the motion to them. In this way motion of the center atom would be transmitted to the eight neighboring atoms in the lattice and propagated through the structure. A laser would be the result of the center atom's oscillation along the diagonal, and all of the motion would be transmitted to four atoms. In the laser the different atoms oscillating within the lattice structure would not produce interference with the motion of other atoms, and all the energy would be transmitted. The structure of various lattices would depend on the atoms or molecules they are made of, but as long as there is a balance between attracting and repelling forces, electromagnetic waves would be transmitted. This lattice structure of matter would explain the dual thin slit experiment. The electrical field of an electron traveling through a lattice would produce electromagnetic waves that would propagate in the lattice.

The state of a material, whether it is a liquid, gas or solid, would depend on the strength of its magnetic and electrical fields. If the repelling force of the electrons were greater than the attractive force of the magnetic field, the structure would be a gas. If the electrical and magnetic forces were equal, it would be a liquid, while a magnetic force stronger than the electrical force would result in a solid. It is the combination of attractive and repelling forces that provides the tension that determines the speed of an electromagnetic wave in a medium, so in the stronger magnetic fields of solids and liquids, light would travel faster than in a gas. The stronger the magnetic attraction, the closer the layers of atoms are together, so the motion of an atom is transmitted faster to the neighboring atoms. In deep space, where the magnetic field is weaker, the electrical force between the elemental hydrogen atoms results in a greater distance between the layers of atoms and a slower transmission of light. This directly contradicts Einstein's assertion the speed of light is constant and that nothing can exceed the speed of light in a vacuum.

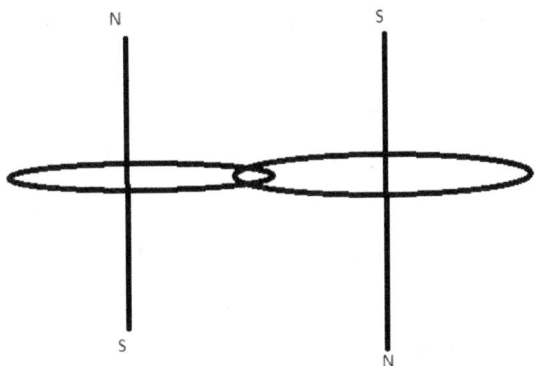

The attractive forces between atoms would also explain the formation of covalent chemical bonds. If an atom in a lattice were missing two electrons in its outer orbit and a neighboring atom had two electrons in its outer orbit, then if one of the atoms were to flip, the electron orbits of the two atoms could mesh like gears, allowing the two magnetic poles of the atoms to come closer together and form a stronger attractive bond than the bond between the atoms in the lattice. The current theory of covalent bonds is that the atoms share electrons, but there is no electrical reason that this sharing should occur.

Helium provides a supporting argument for covalent bonds' being a result of a magnetic attraction between atoms and the meshing of their electrons. Helium has two orbiting electrons, which fills its electron orbit. According to the shared electron theory, there is no need to share electrons, and helium should be completely inert. The fact that helium gas occurs as a diatomic molecule where two helium atoms are joined to form a molecule indicates that helium is not inert and forms covalent bonds. In the meshing gear theory, the small size of helium would mean the magnetic poles of the two atoms would be closer together than in larger atoms and form a strong magnetic bond, thus making helium the most reactive of the elements instead of an inert element. If this were the case, why don't we see helium combine with other elements the way hydrogen does? Hydrogen has only one electron, so it can mesh with many other elements, but helium has two, making it hard for it to combine with larger elements because of the different speed of the electrons in the outer orbits of the two atoms. The orbits would not mesh because of the difference in the energy of the orbiting electrons.

This may offer a reason that helium does not form bonds with larger atoms but does not explain why helium doesn't combine with hydrogen. The two elements may actually combine, but a molecule of helium dihydrogen would be difficult to distinguish from a helium molecule. They would have the same molecular weight and inert chemical properties. The only way to distinguish between the two molecules would be that the attractive magnetic structure of helium dihydrogen would differ from that of two helium atoms. There is an experiment that supports the existence of helium dihydrogen. If you cool helium gas to near absolute zero, some of the liquid helium defies gravity and climbs out of the container. This is presently known as helium one, and it indicates there are different types of molecules in helium gas. Since helium dihydrogen would have a different magnetic footprint than a helium molecule, it is a good candidate to replace helium one.

The atom is made up of two components: matter, which gives it substance, and energy, which gives it structure. Matter gives the atom an electrical field and force, while energy gives it a magnetic field and force. The magnetic forces of atoms then combine to form larger and larger objects. If the current theory of the atom was correct, the nucleus would be surrounded by electrons, forming a shell of negative charges around the atom. This shell would produce a repelling force between all atoms and molecules, so there would be no friction or adhesion, with the exception of chemical bonds. In the theory of the atom with directional electrical and magnetic forces, molecules and atoms could align their forces to provide friction and adhesion without forming chemical bonds. Water, with its small molecular size and non-rigid structure, can act as a lubricant, reducing friction, or as a adhesive, increasing adhesion by modifying the attractive forces between atoms.

The theory that atoms have both repelling and attracting fields is supported by experiments. When graphite is placed over a strong magnet, it floats above the magnet. It does not matter if it is a magnetic north or south pole, indicating that there is a repelling force other than a magnetic force in the structure of the graphite that is causing it to float. The superconductivity experiments support the contention that this repelling force is electrical. In those experiments, exotic ceramics are cooled to close to absolute zero and then placed over a magnet and given an electrical charge. This causes the ceramic disc to float above the magnet, defying gravity. The accepted explanation for the experiment is that when the ceramic disc is cooled to close to absolute zero, electrical resistance in it disappears, and when electrons are added to the disc they create currents that produce a repelling magnetic field, causing the disc to float. This explanation has two problems: Why does this occur in a non-conducting ceramic and not in other materials that have lower initial electrical resistance, and why do the added electrons produce currents rather than disperse to form a charge? If the electrons were to produce currents, according to the right-hand rule, in the presence of a magnetic field the current should flow so as to produce a magnetic field attracted to the base magnet rather than repelled by it. An alternate explanation is that when you cool the disc, you take away energy and weaken the magnetic fields of the atoms in the disc. When the disc is placed over a magnet, you eliminate some of the attractive magnetic forces and replace them with repelling magnetic forces and then, by adding electrons, increase the electrical repelling force, causing the disc to float. You need exotic metals in the ceramic to create the right structure to space the magnetic fields of the atoms, and because the ceramic is a solid the atoms cannot change position to adjust to changing fields.

By having atoms and larger objects being made up of matter with its electrical force and energy with its force, it is possible to explain the phenomena of gravity without creating a force of gravity. First it is necessary to discuss what is wrong with the current theories of gravity.

CHAPTER 7

The Lowdown on Gravity

THE LAW OF gravity seems reasonable and simple. Objects fall because a force is pulling them toward the center of the earth, and the planets orbit the sun because the same force is pulling them toward the sun and preventing their momentum from propelling them out into space. The more you study gravity, however, the more contradictory and unreasonable it becomes, but because it correctly predicts how objects fall and the orbits of planets and satellites it is hard to dispute. The reason Newton's theory gives the correct answer for the orbits of the planets is that to determine the mass of the planets, the law of gravity is used. By equating the force of gravity between a planet and its satellite, the mass of the planet is determined. If you use the theory to get your data, the theory will always give the expected answer.

When the astronauts landed on the moon they discovered it was made not of green cheese but of rocks and minerals, the same as on the earth. Calculations using the theory of gravity to determine the moon's mass show that the moon will have a significantly lower density than the earth (60%), but the data from its exploration shows that this is not the case, and because the moon has no water it should have a higher density than the earth. The probes launched to Mars show a solid planet containing a large amount of iron that use to have active volcanoes, but according to the theory of gravity Mars should be less dense than earth (70%) even though it also has no water. Beyond Mars is the asteroid belt, where the asteroids are composed of metal and heavy elements (as determined by observation) and are denser than the earth, but beyond the asteroid belt is Jupiter, which the law of gravity says is a gas giant one-quarter the density of the earth. The impact of the Shoemaker-Levy 9 comet fragments sent plumes of debris high into Jupiter's atmosphere. If the fragments were striking a gas or liquid at their entry angle, their energy would primarily transfer into waves rather than the debris plumes produced by the comet fragments. The propelling of all that matter and energy into the atmosphere of Jupiter is evidence that the fragments hit a solid planet and not a gas giant. The red color of Jupiter and large magnetic field indicate that Jupiter also contains a large quantity of iron and that the mass determined by gravity is incorrect. This data from direct observations provides strong evidence that the law of gravity is wrong.

Gravity as a force has its own peculiar workings that are unlike the other forces or phenomena produced by the sun. Light from the sun decreases uniformly with distance, resulting in the intensity of light being a function of the distance, and the same is true of the electrical and magnetic forces produced by the sun. But this is not the case with gravity. The sun does not produce a gravitational force, it produces thousands of them. For each object in orbit around the sun, there is a different gravitational force that is specific to that object. The individual gravitational force produced by the sun or a planet is unaffected by other gravitational forces that the same mass produces. Imagine gravity is keeping the planets from falling into the sun instead of pulling them towards the sun, and then picture the force of gravity as a stream of water. The amount of water in the stream would depend on the size of the planet or object being pushed away, and the pressure of the stream of water would depend on how far the object is from the sun. Jupiter, a huge planet at a far distance, would require a large stream of water at high pressure to keep it from falling into the sun, while Mercury, a small planet close to the sun, would require a small, low-pressure stream of water. As Mercury orbits the sun, somehow it passes through the huge stream of water traveling between Jupiter and the sun but is unaffected by it and remains in its own orbit with its own force of gravity.

During a lunar eclipse, the moon is directly behind the earth, which blocks sunlight from hitting the moon, but the force of gravity between the moon and sun is unaffected. This means the moon produces one stream of gravity or water directed at the earth and a larger stream at a higher pressure going through the earth to the sun, and even though these two streams follow the same path they are each independent and unaffected by the other. The sun also produces two different forces of gravity traveling the same path to hold the earth and moon in orbit. The same lack of interaction is true for all the gravitational forces between the sun, earth and moon where the positions of the masses do not affect the independent forces.

The relationship between gravity and energy is even more contradictory and puzzling. When the earth and the moon pull on each other, they do work, which involves the expenditure of energy, but after a month of expending energy they return to the same position with the same energy. You cannot expend energy at 100 percent efficiency, so there must be a loss of energy as heat or in another form, which means the force of gravity violates either conservation of energy or the laws or thermodynamics. The conservation of energy also presents problems for falling objects. As an object falls it increases in kinetic energy, and when it strikes the earth this energy is then transferred to the earth as heat or kinetic energy to other displaced matter. In order for this not to violate the conservation of energy, potential energy was created, where this energy was stored as mass in the falling object and by some unknown means converted into kinetic energy as it falls. This contrivance may offer an explanation for falling objects in the earth's gravitational field, where there is only one target, but when you apply it to meteors and asteroids that are not in the earth's gravitational field it presents problems. An object in space can hit not only the earth but also the moon, Mars, Jupiter or any of the other objects orbiting the sun. The kinetic energy for each of these collisions is different, so the energy stored as mass must take into consideration all of these possible collisions. As an asteroid heads on a collision course with the earth, does its mass decrease as potential energy is converted into kinetic energy or does its mass increase as the potential energy between it and Jupiter and other objects increases? As the planets and asteroids orbit the sun, the distance between the asteroids and various targets changes, which means that the potential energy also changes and that their mass would be continually changing as they orbit the sun. If the mass is changing, doesn't this necessitate a change in the gravitational forces between all the objects orbiting the sun? The solution of having asteroids and meteors knowing the future, like electrons do, is not an acceptable explanation.

According to physicists, the force of gravity is not proportional to the masses that create it. One hundred percent of the gravitational force between the sun and the earth comes from the sun and zero percent comes from the earth, while simultaneously zero percent comes from the sun and 100 percent comes from the earth. The inability to apportion the force is a result of having different forces produced by the same mass for different orbiting objects; same mass has to produce multiple gravitational forces.

A black hole is a mass where the gravitational pull is so strong that even light, with no mass, cannot escape. To preserve the conservation of mass and energy any increase in energy and mass of an object accelerating toward the black hole must come from the black hole. This means that the contradictory assignment of force of gravity between objects is not always true and that there should be a way to apportion the force according to mass. The fact that this cannot be done is another failure of the theory of gravity.

Perhaps the most persistent problem with gravity is the tides on earth. It is clear that the tides are a function of the moon's orbiting of the earth, but the moon has a consistent orbit and the tides are anything but consistent. In some places high tide occurs only once a day instead of twice, while in other areas whether it occurs or not is unpredictable. Then there is the question of why does the moon produce a high tide on the far side of the earth? The first explanation was that the moon was pulling the earth away from the water on the far side of the earth and causing a high tide, but this is ridiculous. Using vectors to show the gravitational forces on an object on the far side of the earth and on the near side, you would start with two large arrows pointing toward the center of the earth representing the gravitational pull of the earth on the objects. The vectors for the gravitational pull of the moon on these objects would then be added to these vectors. The moon's pull on the object on the far side of the earth would be in the same direction as the vector from the earth's gravitational pull and so would increase the length of that vector. The vector representing the moon's force of gravity on the object on the near side of the earth would be in the opposite direction as the earth's gravitational pull and would be subtracted from that vector. The water on the far side of the earth has a stronger gravitational force pulling it toward the center of the earth than the object nearest the moon, and according to the vector analysis the moon should cause a low tide, not a high tide, on the far side of the earth.

Since the gravitational force between the sun and earth is even greater than the force between the moon and the earth, why doesn't the sun cause a high tide at noon and midnight? The fact that the gravitational pull of the sun does not affect the tides except in conjunction with the moon's gravitational pull would indicate that something is wrong with having the force of gravity being the cause of the tides or that the current theories of gravity are wrong.

These problems with gravity are not the ones that bothered Einstein and led to his theories of gravity or general relativity. The light hitting the earth does not leave the sun directed at the earth but directed in front of the earth – it takes time for the light to travel all that distance. However, according to Newton the force of gravity is instantaneous, acting like a string between the sun and earth, holding the earth in its orbit. If that string were to break, the earth would immediately proceed on a straight line into space. Einstein's objection was that gravity travels faster than the speed of light, and he maintained that nothing can travel faster than the speed of light. His theory of general relativity was developed to have gravity travel at the speed light, and according to it mass distorts the space-time continuum, causing the planets to travel in orbits. The prediction of the bending of light around the sun, the shift in light of distant stars and a more accurate prediction of Mercury's perihelion (the progression of the apex of Mercury's orbit) were seen as confirmation of Einstein's theory, but there needs to be an examination of other possible causes for these phenomena. Light is an electromagnetic wave, and the reason light bends around the sun is that the stronger magnetic field closer to the sun causes a change in the speed of light. The varying strength of magnetic fields in space would also cause the red and blue shifts, as they changed the speed of light. As for the cause to the change to Mercury's perihelion, it remains unknown. The bending of the four-dimensional space-time continuum necessitates the creation of a fifth dimension, just as taking a two-dimensional piece of paper and bending it creates a third dimension. The creation of new dimensions to solve problems of theory abandons the principle that theory should be based on observations and instead makes reality conform to theory. It is odd that in special relativity the speed of light is a constant and distorts mass and time, while in general relativity mass and time distort light.

A major problem with both theories of gravity is that there is not enough mass to hold the universe together and to hold the solar systems at the outer edges of galaxies in the galaxies. This problem resulted in the creation of dark matter and dark energy. Today physicists believe that over 90 percent of the matter in the universe is dark matter and is invisible. Physics is a science that is supposed to be based on observations of reality; instead of inventing imaginary solutions to problems physicists should consider alternate explanations based on observed phenomena. The mistaken belief in the expansion of the universe is a result of thinking that the red shift of light is a result of a Doppler effect caused by an expanding universe rather than the result of the electromagnetic waves traveling through magnetic fields of different strengths. The problem that there is not enough matter in a galaxy to prevent the outer solar systems from flying out into space also needs an explanation that is based in reality. What is needed is a new theory of gravity that will rectify the deficiencies in the present theories without the creation of new dimensions or invisible entities.

CHAPTER 8

Gravity X Mass

WHEN NEWTON PROPOSED his theory of gravity, he made an assumption that an object in motion will continue in a straight line unless a force acts upon it. Nothing, not even light, travels in a straight line, so how can you know what a straight line is? The new theory of gravity starts with a different assumption: An object in motion will maintain its energy unless energy is added to it or given off by it. It also asserts that energy is not a property of an object or a different form of matter but a thing, one of the fundamental building blocks of the universe. Energy and matter combine to form units where the matter provides substance while the energy produces structure. The theory then asserts that energy fields can combine to form larger and larger units, and finally it proposes that a unit within an energy field will equalize with the energy field by gaining or losing energy to the field.

The reason the asteroids between Mars and Jupiter did not become a planet is that Jupiter's large energy field overlaps the asteroid belt and the asteroids orbiting the sun are trying to equalize with two different energy fields, Jupiter's and the sun's, at different times. It is interesting that within the asteroid belt some asteroids are observed orbiting other asteroids. This behavior cannot be explained by a force resulting from the mass of the asteroids but can be attributed to the energy field of an asteroid.

The rate at which an object falls and the speed of an object in orbit have nothing to do with mass, so why would these actions be a function of mass? The phenomena we attribute to gravity have nothing to do with mass but are a result of energy. The motion of objects either falling or orbiting is a result of the interaction of energy fields of different units instead of a force associated with the mass of the units. The earth and moon are each units with energy and electrical fields. The two energy fields combine to form the earth-moon unit, and this is the unit that orbits the sun. The energy field of the sun radiates out from the sun just like light and decreases according to distance. If you take the orbital energy of a planet, velocity squared, times its distance from the sun, you will get the same value for all the planets, and this represents the energy of the solar system unit. The planet units orbit the sun because they are in equilibrium with the energy level of the sun at that distance, not because they are being pulled toward the sun. They are coasting, and to move into a different orbit they must have a change of energy, either gaining or losing energy. If you take the velocity squared of a satellite orbiting earth times its distance from the earth, you have the energy of the earth unit. If you add energy to a satellite, increasing its speed, it is no longer in equilibrium and moves into a weaker energy field farther from the earth, losing energy to the field, resulting in it orbiting slower or having less energy than when it started. If you take energy away from a satellite by slowing it down, it then moves into a higher energy field, closer to the earth, gaining energy from the field, and ends up going faster or having more energy than when it started. It would seem counterintuitive that to make satellites go faster you step on the brake while to make them go slower you step on the gas, but that is how orbital dynamics work. An object will equalize with the energy field it is in by gaining or losing energy to the field.

When an object falls, it is not being pulled towards the earth, it is moving into a denser energy field and gathering energy. The kinetic energy of a falling object increases because it is gaining energy as it moves into higher energy fields, and if an object gains enough energy or speed it will equalize, going into orbit. Scientists use this to increase the speed of satellites going to other planets. By having a satellite pass close to a planet, it gains energy and increases in speed, adding speed for its continued journey. If there was a force of gravity pulling on the satellite, for every position moving toward the planet where the force of gravity is adding to its speed there is a corresponding position when it moves away from the planet where gravity is slowing it down. The reason the satellite gains energy is that it is unable to radiate energy into the field fast enough and leaves the field before losing all the energy it gained.

That the moon causes a tide while the sun does not has to do with the combining of energy fields. The individual energy fields of the earth and moon combine to form a larger energy unit that orbits the sun, and the combination of these two energy fields results in an increased energy field along the axis between the two units. On the earth, water will try to equalize with the increased energy along the axis and will fall up to form the high tides on either side of the earth. Because the sun is not pulling on either the moon or the earth and the combined earth-moon unit is in equilibrium with the sun's energy field, the only time the sun affects the tides is when the sun's energy field aligns with the earth-moon energy field.

The velocity of a planet's moon squared times the distance from the planet gives the energy of that planet, and this energy unit is what orbits the sun. All objects at a given distance from an energy center will travel at the same velocity, and the mass has nothing to do with the velocity; this is why when momentum and the force of gravity are used to calculate the mass of a planet the result does not agree with the observed data.

The sun and planets form a unit, the solar system, which then combines its energy fields with the energy fields of other solar systems, forming the galaxy, and galaxies' fields in turn combine to form groups of galaxies. Fields of energy and matter combine to form larger units with their own organizing energy fields. The unexpected bulging shape of the bullet galaxy is the result of the fields of two galaxies colliding, not a result of dark matter. If the gravitational pull of dark matter was involved, the galaxy would bulge in both directions. There is no need to invent invisible dark matter and dark energy to explain how the universe works; it is a function of the fields of various units.

CHAPTER 9

How It All Works

WHY AND HOW do energy and matter interact? Why is an alpha particle, which is electrically unstable, stable while a neutron molecule, which is electrically stable, unstable? To answer these questions we need to look at radioactive atoms and discover why certain isotopes are unstable. If the nucleus of an atom has too many or too few neutrons or electrons, it is unstable resulting in it either splitting or emitting particles to form more stable structures. A neutron and alpha particles are not particles but a hydrogen atom and helium atoms splitting from the nucleus. The beta radiation of a nucleus is a particle emission of an electron by the nucleus. If the electron acts as glue holding the protons together in the nucleus, how can the nucleus emit it? How does it break away from the neutron molecule in the atom's nucleus, overcome the attractive force of all the protons in the nucleus and get propelled through the repelling force of the orbiting electrons with such force?

For the electron to overcome all the forces holding it in the nucleus, a large amount of energy must be acting on it, and this provides a clue to how energy and matter interact. Energy is attracted to positive matter, protons, and repels negative matter, electrons. The strong nuclear force holding the positive charges of the protons together is not a binding force but a compression force produced by the energy being attracted to the protons in the nucleus and repelling the orbiting electrons. If the nucleus has an electron exposed to the energy force, as in a neutron molecule, energy -- being a stronger force than the electrical force -- will dislodge the electron and expel it from the nucleus or split a neutron molecule. The alpha particle with four protons and two electrons is stable because the electrons are in the center and surrounded by protons, and are not exposed to an energy force. An atom is radioactive emitting particles or splitting into more stable atoms. If there are too many neutrons or electrons in the nucleus so they cannot be shielded by protons the force of energy will destroy the nucleus. If there are too few neutrons or electrons there will not be enough electrical or weak force to hold the nucleus together and the force of energy will be able to destroy it. The force of energy is a stronger force than the electrical force of matter (probably by a factor of psi, and this is why the universal constant keeps occurring in the orbits of electrons, planets and throughout nature). By having energy be attracted to protons with a stronger force than the electrical attraction between electrons and protons, we eliminate the nuclear forces, eliminate gravity and split the electromagnetic force into two forces, one a product of energy and the other a product of matter. It is the interaction of the two fields that causes motion and builds the structures that make up the universe.

An interesting result of this theory is that according to it the sun burns backward. The sun starts as a mass of neutron molecules, which is the base state of matter. The positive charge of the protons attracts energy, which then tries to dislodge the electron and convert the neutrons into a hydrogen atom, hydrogen having the highest energy-to-mass concentration of any element. In doing so energy breaks off larger chunks of neutron molecules from the neutron star and then attacks these clumps. The force of energy is able to break up the clumps until the electrons of the nucleus are shielded by protons and the clumps form stable atoms.

This secondary conversion of clumps of neutrons to atoms may be the reason the corona of the sun is hotter than the surface of the sun. The stable atoms of hydrogen and helium are the waste products or ash of the reaction of the sun and not the fuel. The larger elements are also waste products, and when the concentration of ash and waste gets high enough that it impedes the energy from getting to the neutrons, the sun explodes, propelling the ash into space, where it congeals and becomes the trash piles called the planets and moons. The current theory that says the heavier elements found on earth are the product of far-away exploding stars that have gone nova doesn't make sense. The mass of the solar system is miniscule compared to its volume, and if the sun exploded, propelling all the matter in the solar system into space the amount reaching the closest star would be infintisimal. If the size of the solar system (radius equal to the outer edge of the Oort cloud) and its mass (overestimated as twice the mass of the sun) was shrunk to the size of the earth its mass would be less than three hundred grams. To believe that the tons of heavy elements that occur on the earth and other planets came from other galaxies or even from within the Milky Way is not credible because it would require a concentration of matter in space that does not occur. It is called empty space for a reason. The chemical elements that make up our solar system originated in our sun. When a sun explodes for the first time, it creates a solar system, and subsequent explosions then add matter to the trash piles.

CHAPTER 10

Conclusion

BY SEPARATING THE electromagnetic force into two forces, one produced by matter and the other by energy, it is possible to eliminate the forces of gravity and the strong and weak nuclear forces. The different strengths of these forces and their interaction allows for the building of the atom and the subsequent creation of larger units without necessitating the creation of a myriad of subsubatomic particles and dimensions. Physics once again conforms to reality and reason.

Appendix 1

Experiment to determine the force between magnets

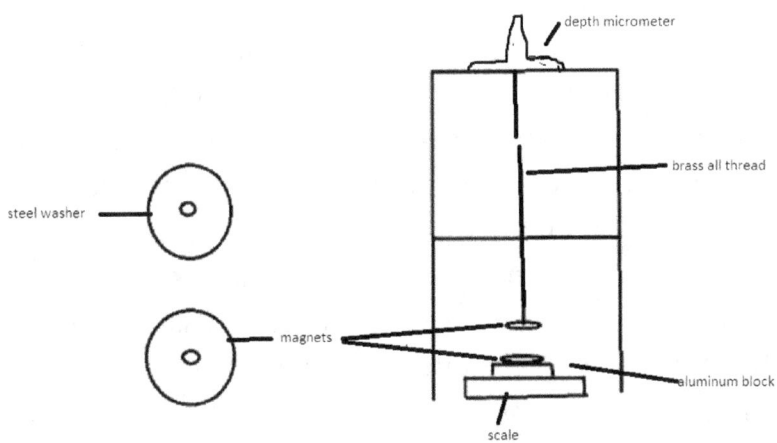

After deciding that the law of gravity was wrong I decided to test whether the similar formula for the force between magnets was also wrong. The experiment consists of testing the strength of two individual magnets and then measuring the force between them.

The equipment used consistent of a block " A " frame made of a 2-inch aluminum angle with a slot cut in the middle cross bar on which to hang a 1/8 brass all thread rod by a brass nut. On the top crossbeam a hole was drilled over the slot so a depth micrometer could be used to measure the change in distance as the rod is lowered. The magnets were two round composite magnets with a center hole in which a brass nut, which had been ground to fit, was glued. A steel washer with a similar brass nut glued in the center was also used.

To start the experiment a steel block was placed on a scale under the "A" frame and the tare of the scale was set to zero. The length of the rod was then adjusted so the bottom of the rod was just above the steel block and the scale still read zero. The distance from the top of the brass rod to the top of the "A" frame was then measured by the depth micrometer, giving the zero distance between the magnet and steel block. One of the magnets was then screwed onto the all thread so the bottom of the magnet was flush to the bottom of the rod. The hanging nut was then screwed down the rod to raise the magnet from the zero distance, and using the depth micrometer the distance between the magnet and steel block was obtained and a value for the force of the magnet was measured in grams.

The nut was then screwed up the rod, lowering the magnet, and at various intervals measurements were made of the strength of the magnet at the distance to the steel block. This process was repeated with the second magnet to determine its strength at various distances. The results of the two magnets were then plotted on graph paper, producing the typical curve that shows the magnet strength decreasing by approximately the cube of the distance. By projecting the curves to a zero distance, the strength of each magnet was found.

One of the magnets was then attached to an aluminum

block by a brass screw threaded through a tapped countersunk hole. The end of the screw was adjusted to be flush with the top of the magnet while holding the magnet tightly to the aluminum block. The procedure for finding the strength of the individual magnets was repeated to determine the force between the two magnets, first determining the zero distance for the brass rod to the top of the magnet on the aluminum block, then screwing the other magnet onto the rod and measuring the force at various distances. The results were plotted on graph paper, producing a curve that looked as if the force between the magnets decreased according to the square of the distance.

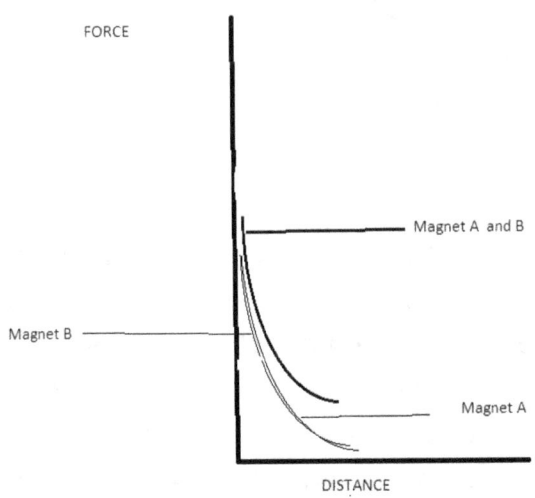

FORCE OF MAGNETS AND BETWEEN MAGNETS

The odd thing with this result was that the force between the two magnets at the zero distance was not very different than the force of a single magnet at the zero distance.

This led to the conclusion that the measurement of the strength of the single magnets also captured the strength of an induced magnet in the steel block and therefore was not accurate. As the magnet descended, the magnetic field rearranged the structure of the steel block, releasing a magnetic force from the steel block. The magnetic force of the permanent magnet increased as it descended, and this increased the strength of the induced magnet, resulting in a measurement that was the force between two magnets, a constant permanent magnet and a varying induced one. The varying magnet's force approached that of the permanent magnet, so at the zero distance they were almost equal, giving the same result as the force between the two permanent magnets. Even if the strength of the magnet was half of the measured strength, the force between the two magnets was nowhere near the expected product of the magnets.

ﮞﮞﮞﮞ

This led to considering how the distance between the magnets was measured. The traditional measurement is from the center of one magnet to the center of the other, but this would mean that the magnetic force is decreasing within the body of the permanent magnet. This makes no sense, since the magnetic material on either side or the midpoint adds to the strength of the magnetic field. It seems more reasonable to believe that the magnetic force is constant within the magnet and begins to decrease only upon leaving the magnet. To test whether this was the case, the brass rod was lowered to just above the magnet on the aluminum block and secured to the frame with brass nuts. A brass jam nut was screwed up the brass rod a short distance, then the other magnet was tightened against the jam nut, securing it. The steel washer was then screwed up the rod against the magnet, creating a magnet consisting of the two permanent magnets and the induced magnet of the steel washer, with a space between the two ends of the magnet. If the strength within a magnet was constant, as the steel washer descended the rod the reading on the scale should stay constant. The results were that as the washer descended, the reading on the scale was constant until the washer reached the midpoint between the magnets, at which point the force began to increase. As the washer descended further toward the bottom magnet, the force increased exponentially.

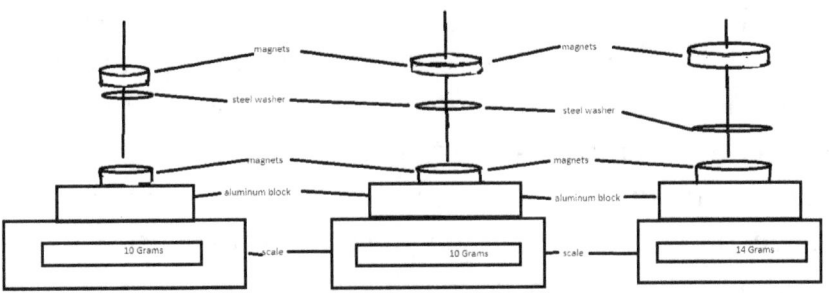

The result indicates two things. First, the force of the magnet is decreasing linearly with distance in the first half of the descent because if it decreased as a cube of the distance the strength of the induced magnet would decrease, resulting in a decline in the force of the combined magnets. Second, the distance between two magnets is from the face of one magnet to the magnetic field of the second magnet. When the washer leaves the magnetic field of the upper magnet at the midpoint, the bottom magnet is then attracted to it, resulting in the reading of an increasing force, but while the washer is in the magnetic field of the upper washer it is invisible to the base magnet. This would account for the fact that the measurement of the strength of the single magnets approximated that of a cube function. Not only was the strength of the induced magnet in the steel block increasing as the permanent magnet approached it, but the size of its magnetic field was also increasing, resulting in the distance between the magnet and the steel block decreasing by more than the measured amount.

The strength of the two individual magnets was almost equal, so by calling them equal a new formula for the force between two magnets, $F = 2m/d/2$, could be compared to the traditional one, $F = m$ squared$/d$ squared. By using the values for the force at different distance we could solve each equation for the strength of the magnet, which should be constant. In the new formula the distance is the distance between the faces of the magnets, while in the traditional one the distance is between the centers of the two magnets. The two formulas, $m = Fd/4$ and $m =$ square root of F times d squared, gave similar results where the value of m was fairly constant until the distance between the magnets became small, and then the value of m varied significantly in both formulas.

If both formulas give similar results, the correct formula would be the one that best describes what is occurring during the experiment. Two squared gives the same value as two plus two, but they are different actions. The experiment starts with two independent magnets that have their magnetic flux lines circling around them. As the magnets get closer, their flux lines combine, forming a third magnet, and it is the strength of this magnet that we are measuring. The closer the magnets get, the more flux lines combine, and the third magnet becomes stronger as the two individual magnets get weaker. When the magnets get close enough, all the flux lines are combined, and then the subsequent increase in strength is from the shortening of the flux lines. The new formula gives a better description of what is happening where the forces of the two magnets are added together to form a third magnet. The results show that magnetic forces combine to create larger magnets.

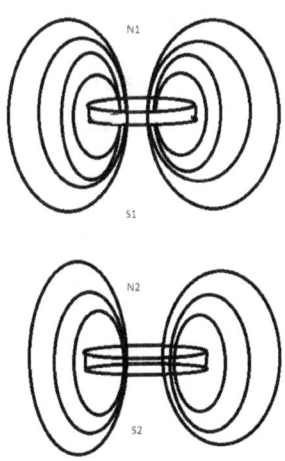 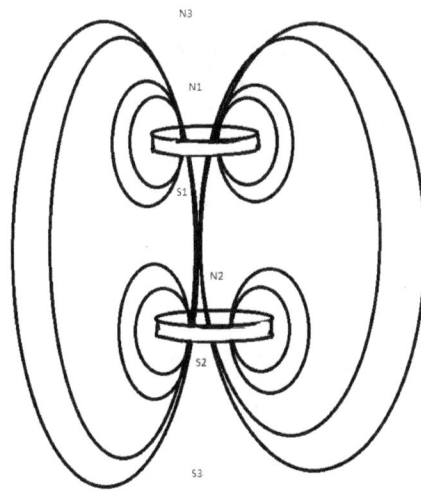